A book
is a present you can open
again and again.

THIS BOOK BELONGS TO

_____

FROM

_____

MAD ABOUT MATH

Illustrated by Rick Incrocci

TREASURE TREE™

World Book, Inc.
a Scott Fetzer company
Chicago London Sydney Toronto

Dear Parent,

You may want to join your child in making some of the projects in this book, or be present in the room where your child is working. If you are with your child, you can lend a hand, answer questions that may arise, and make sure that your child is using materials appropriately.

Have fun!
The Editors of World Book

Printed in the United States of America
ISBN 0-7166-1618-1
Library of Congress Catalog Card No. 91-65751

D/IC

Cover design by Rosa Cabrera

Pattern Sticks

You will need:

wooden skewers
fruit cut into bite-sized
   pieces
small marshmallows
paper
pencil

Pattern sticks are fun to make and taste good, too. You can use all kinds of fruit to make your sticks: banana slices, apple wedges, orange slices, pineapple chunks, pitted cherries, grapes, and melon balls.

1. Wash your hands carefully. Spear the pieces of fruit and the marshmallows onto skewers in whatever order you like. You might want to design your patterns with attention to size, shape, or how you think the combinations of fruit will taste.

2. Remove the fruit and marshmallows from a pattern stick and eat them. Sample the other pattern sticks.

3. Keep a record of your best-tasting recipes. Make a Pattern Stick Recipe Book.

Sorting It Out

Putting things that belong together in the same place is called sorting. Sorting can be helpful. Ask your parents if you can sort some of the things in your house.

Laundry

1. Sort your clean clothes in piles (a pile for socks, a pile for shirts, and so on).
2. Sort the clothes according to the drawer or shelf they belong in.
3. Put your clothes away where they belong.
4. Is anything left? What did you forget to sort?

Kitchen

1. Sort everything on one shelf in a kitchen cupboard. Try sorting by kind of container (jar, can, box). By kind of food. By size. By weight. By color. By smell.
2. Help to empty grocery bags. Sort the groceries according to where they should go (refrigerator, cupboard, freezer).

Amazing Fact

The United States makes more glass and plastic containers than any other country. Each year, it makes about 46,000,000,000 glass containers and about 11,000,000,000 plastic ones.

Peanut Butter Fudge Balls

Ask a grown-up to watch you make this treat.

You will need:
measuring cups
peanut butter
honey
mixing bowl
mixing spoon
powdered milk
wax paper

1. Measure 1 cup (.24 liter) peanut butter and 1 cup (.24 liter) honey.
2. Mix the peanut butter and honey together in a bowl.
3. Measure $1\frac{1}{2}$ cups (.36 liter) dry powdered milk.
4. Add the powdered milk to the peanut butter and honey mixture. Mix some more.
5. Roll bits of the fudge on a piece of wax paper to form little balls.

The recipe makes 1 dozen peanut butter fudge balls, so you'll have plenty to share.

Straw Mobiles

1. Thread a long piece of string through three straws. Leave some string hanging out at both ends.
2. Tie the ends of the string together so that the straws form a triangle.

3. Make a loop at the top to hang the triangle.
4. Cut a straw in half. Use one half and two whole straws to make another triangle.
5. Cut straws into different lengths and make more triangles.
6. Ask a grown-up to straighten a wire coat hanger and hang it from a ceiling lamp or hook.
7. Hang the triangles from the wire hanger. (Ask a grown-up to help.) Now you have a mobile.

Mix It: Shape It

In this activity, you can make and bake any shapes that you like. You can hang the shapes or give them away as gifts.

You will need:

measuring cup
mixing bowl
4 cups (.95 liter) flour
1 cup (.24 liter) salt
$1\frac{1}{2}$ cups (.36 liter) water

food coloring (optional)
toothpick, bottle cap, fork (optional)
paper clips (optional)
a grown-up to help
cookie sheet

1. Measure the flour and salt and mix them together well.
2. Measure the water. If you want colored dough, add food coloring to the water.
3. Add the water to the flour and salt, and mix it with your hands.

4. Knead the dough with your hands for four to six minutes until it is smooth.

5. Take small bits of dough and shape them into squares, triangles, circles, and rectangles. The shapes can be thin and flat or thick and rounded.

6. If you like, use a toothpick, bottle cap, fork, or your finger to make designs on your shapes.

7. If you want to hang a shape, here's how to make a hook. Open a paper clip and push it into the top of the shape before baking.

8. Ask a grown-up to heat the oven to 300° F. (149° C) for colored dough or to 350° F. (177° C) for plain dough.

9. Place your shapes on a cookie sheet, and bake for one hour or more until hard.

The Shape of Things

You will need:
construction paper
   (different colors)
scissors
glue or paste
crayons or markers

1. Cut different shapes out of colored construction paper.

2. Paste the shapes on another sheet of paper to make a picture.

3. Finish your picture by adding details with crayons or markers.

4. Give your picture a title.

Tantalizing Tangrams

A tangram is a Chinese puzzle. Tangram pieces can be put together to make many different shapes. Try this activity and see.

You will need:

paper
pencil
scissors

1. Copy the tangram square pattern below onto a piece of paper.
2. Cut along the lines you drew to make seven pieces.

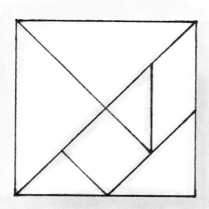

3. Make a house using all seven pieces.
4. Make a boat. See if you can figure how the pieces fit together.
5. Now make your own animal, person, or thing using the seven tangram pieces.

Play Fair and Square

You will need:
one-inch graph paper
scissors
construction paper
glue or paste

1. Cut five 1-inch (2.5-cm) squares from the graph paper.
2. Make a shape with the five squares. Follow this rule: Each square must share a full side with at least one other square.
3. Glue your shape onto a piece of colored paper.

There are 12 shapes in all that you can make with five squares. Cut out more squares, and try to make all 12.

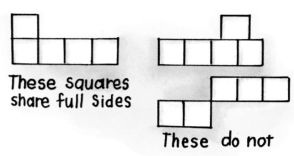

These squares share full sides

These do not

Do you know that there is something that stays in one piece after it is cut in half? In this activity, you will see how.

The Mysterious Möbius Strip

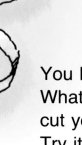

Experiment One

1. Cut off a strip of adding machine tape. Tape the ends together to make a circle.
2. Cut the circle in half down the middle. What do you get? You get two circles, of course!

Experiment Two

1. Cut off another strip of adding machine tape. This time, twist one end before you tape the ends together.
2. Cut this twisted circle in half down the middle. What happens?

You have made what is called a Möbius strip. What do you think would happen if you cut your Möbius strip down the middle again? Try it and find out.

Bored? Not with a Geoboard!

In this activity, you will build a geoboard. You can use your geoboard to make all sorts of shapes and patterns.

You will need:

paper
pencil
ruler
scissors
tape

$7\frac{1}{2}$-inch (18.75-centimeter) square of $\frac{1}{2}$-inch (1.25-centimeter) plywood
25 1-inch (2.5-centimeter) finishing nails
hammer
a grown-up to help
colored rubber bands

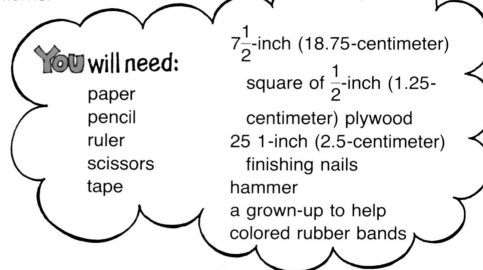

1. Measure and cut out a $7\frac{1}{2}$-inch (18.75-centimeter) square of paper.

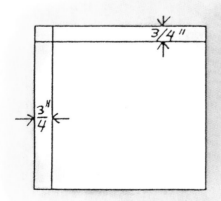

2. Use a ruler to draw two lines on the paper. One line should be $\frac{3}{4}$ inch (1.88 centimeter) from the top, and the other should be the same distance from the left side.

14

3. Draw four more lines across and four more lines down.

The lines should be $1\frac{1}{2}$ inches (3.75 centimeters) apart.

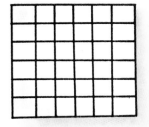

4. Place the paper on top of the wood. Line up the edges. Then tape the paper down.

5. Ask a grown-up to help you hammer a nail wherever two lines cross. Make sure the nails don't go through the bottom of the board.

6. Tear off the paper. Your geoboard is ready to play with.

7. Make a pattern on your geoboard. Stretch a rubber band around some nails to make a triangle. Count how many nails the rubber band touches.

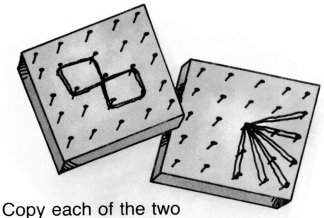

8. Copy each of the two patterns above one at a time on your geoboard. Continue each pattern as far as you can.

1. Stretch a rubber band around 8 nails on your geoboard to make a shape like a stop sign. How many nails are outside the shape? Inside?

2. Make a triangle. How many sides does your triangle have? How many corners does it have? Can you make a large triangle, a small triangle, a tall triangle, a short triangle, and a strange-looking triangle?

3. Make a square. How many sides does it have? How many nails does each side touch? Make the largest square you can. Make the smallest square you can. Make a medium-sized square. Tilt the shape by moving the rubber band over at the top.

4. Make a rectangle. How is it different from a square? How is it the same? Make a tall rectangle. Make a short rectangle. Then tilt the shape.

5. Combine triangles, squares, rectangles, and other shapes to create a pretty, colorful design.

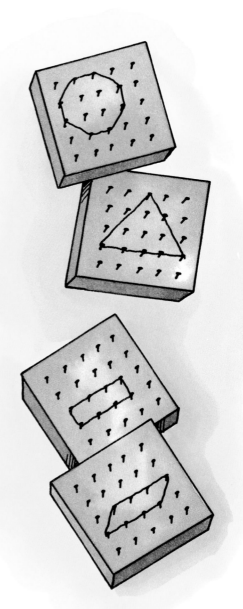

16

You will need:

ruler
paper
pencil
scissors
geoboard
small and large colored
rubber bands

1. Measure and cut out a paper square $1\frac{1}{2}$ inches by $1\frac{1}{2}$ inches (3.75 centimeters by 3.75 centimeters).

2. Stretch a rubber band around four nails on your geoboard to make a square.

3. Fit your paper square inside your geoboard square. We call this shape one square unit.

4. Stretch a rubber band around eight nails to make a larger square. How many paper square units will fit inside it? What fraction of the square is one paper square unit? Two paper square units?

5. Now make a bigger square around 12 nails on your geoboard. How many paper square units will fit inside it? What fraction of the square is one square unit? Two square units?

On The Side of Symmetry

You will need:

paper
pencil
scissors

When two sides of something look the same, we say they are symmetrical. You can use a fold line to make a line of symmetry.

Hearts

1. Fold a piece of paper in half.
2. Draw half a heart at the fold.
3. Cut on the line you drew. *Don't* cut on the fold.
4. Open your shape. What shape have you made? How do the two halves compare?

Can you make a butterfly the same way?

Cutouts

1. Fold a piece of paper in half.
2. Cut out any small shape you like on the fold.
3. On another paper, draw what you think the shape will look like when it is unfolded.
4. Then open your paper and compare.

Try cutting other shapes at the fold and guessing how both parts will look together.

Mirror Mirror

You will need:

magazines
scissors
paper
glue or paste
small mirror
crayons or
   markers

1. Cut out a picture of a large face from a magazine. The face should be looking straight at you.
2. Fold the face in half (along the nose) and then cut it in half on the fold line.
3. Glue half of the face to a piece of paper. Throw the other half away.
4. Hold a mirror straight up and down on the cut line. What do you see?
5. Take away the mirror, and draw in the other side of the face.

Go on a symmetry hunt. Look for objects and pictures around your house that are symmetrical. How many can you find?

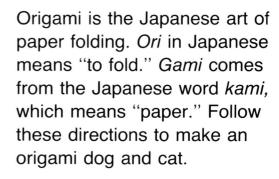

Origami Originals

Origami is the Japanese art of paper folding. *Ori* in Japanese means "to fold." *Gami* comes from the Japanese word *kami,* which means "paper." Follow these directions to make an origami dog and cat.

You will need:
2 square pieces
of thin paper
crayons or markers

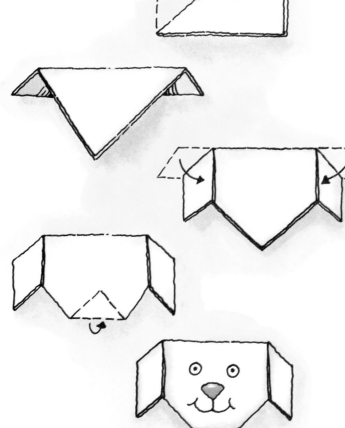

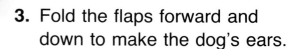

Dog

1. Fold a paper square in half crosswise to make a triangle.

2. Work with the folded edge at the top. Fold this edge backward to make left and right flaps.

3. Fold the flaps forward and down to make the dog's ears.

4. To make the dog's chin, fold the bottom point back.

5. Draw the dog's face.

Cat

1. Fold a paper square in half crosswise to make a triangle.

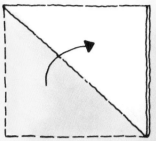

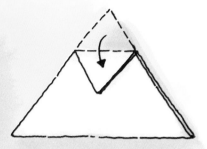

2. Work with the folded edge at the bottom. Fold the top point forward to make a flat top.

3. From the center point at the bottom, fold the two bottom corners up and over the top. Make sure your paper looks like the picture.

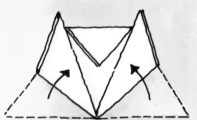

4. Turn your paper over and draw the cat's face.

How Many Feet Do You Need?

Long ago, people measured things by using parts of their bodies. In this activity, you'll use your foot as a unit of measure.

1. Make a footprint by placing your foot on a piece of paper and drawing an outline of it. Cut out your footprint.
2. Find five things that are longer than your footprint. Find five things that are shorter. Find five things that are about the same length.

You will need:

pencil
paper
scissors

I NEED AN AREA RUG ABOUT 20 NOSES BY 30 NOSES.

RUG DEPT.

3. Use your footprint to measure the length of other items. You could measure your mom's arm, your bed, your pillow, or even the bathtub.

4. About how many footprints long is each thing that you measured?

UNIT		
SHORTER	ABOUT THE SAME	LONGER
MUSTARD JAR	ENVELOPE	BASEBALL BAT

THING	LENGTH
MOM'S ARM	
PILLOW	
BED	
BATHTUB	

Crash Landing

If a spaceship made a crash landing on earth, what are the chances of it hitting land? Make a guess. Do you think it would hit land or water? Try this experiment to see.

You will need:
world map
coin

1. Toss the coin so that it lands somewhere on the map.
2. Toss the coin 10 more times.
3. Keep track of where the coin landed each time.
4. Did the coin fall more often on land or in water? Why do you think this is so?

This is a game that is played in Spain. Read the directions. Then teach your friends or family how to play.

1. All players form a circle facing each other with their hands behind their backs. Each player holds three beans in his or her hands.

2. Carefully, so the others don't see, each player puts zero, one, two, or three beans in the right hand.

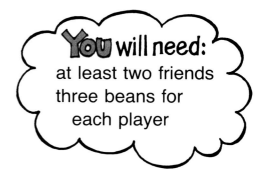

You will need:
at least two friends
three beans for
each player

3. All players then put their closed right hands in the middle of the circle.

4. Taking turns, each player tries to guess the total number of beans held in the circle. Players aren't allowed to guess the same number.

5. After each player has had one turn, the players open their hands and count the beans.
6. If a player has guessed the total number correctly, he or she gets a point.
7. The first player to guess correctly three times wins the game.
   Hint: Listen carefully to the guesses.

Try playing this game with four or five beans each.

PLAYER	TALLY
SUE	III
JOE	
DAN	II

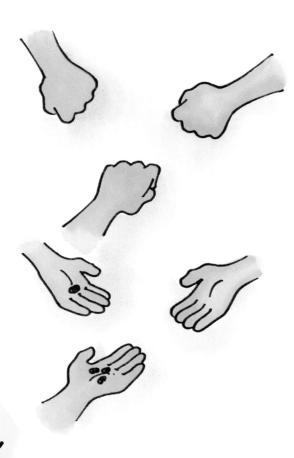

⭐ **Amazing Fact**

Several hundred years ago, large silver Spanish dollars were called pieces of eight. To make change, a person could chop the coin into eight pie-shaped pieces called bits. Two bits were worth a quarter of a dollar, four bits a half dollar, and so on. The saying "two bits" still is used to mean 25 cents.

Kitchen Concert

You will need:
eight glass bottles
water
spoon

Line up eight bottles that are the same size. Leave one empty. Then, by estimating, fill the other seven bottles one-eighth full, one-fourth full, one-third full, one-half full, two-thirds full, three-fourths full, and completely full.

Tap the bottles with a spoon. How do they sound? Can you play a song?

empty $\frac{1}{8}$ $\frac{1}{4}$ $\frac{1}{3}$ $\frac{1}{2}$ $\frac{2}{3}$ $\frac{3}{4}$ FUll

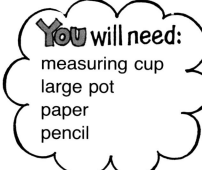

You will need:
measuring cup
large pot
paper
pencil

1. Look at the pot and estimate how many cupfuls of water it will hold.
2. Write down your estimate.
3. Now use the measuring cup to fill the pot. Keep track of how many cups you pour into the pot.
4. How close was your estimate?

Here are some other things to estimate.
How many cups of juice will fill a pitcher?
How many grains of rice in a teaspoon?
How many slices in a loaf of bread?
How many crackers or cookies in a box?
How many oranges, potatoes, or apples in a bag?

How Much Is One Hundred?

Make a collection of things that amount to 100. Label your collection. For example, you could count and collect 100 marbles, 100 pebbles, 100 paper clips, 100 pennies, 100 rubber bands, 100 macaroni shells, 100 toothpicks, 100 acorns, or 100 buttons. How many different collections of 100 things can you make?

People have been keeping time for thousands of years, but they have not always had the same kinds of clocks that we have today. Here's an easy way to keep time.

The Sands of Time

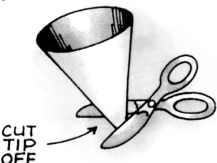

CUT TIP OFF

1. Cut a small hole in the bottom of a cone-shaped paper cup.
2. Place your finger over the hole and pour sand into the cup.
3. Hold the paper cup over the mouth of a jar and remove your finger from the hole. Let the sand flow into the jar for one minute.
4. At the end of one minute, place your finger over the hole in the cup and throw away any leftover sand.
5. Use the cup, the sand, and the jar as a one-minute timer.

Can you make a two- or five-minute timer?

You will need:
cone-shaped paper cup
  or a coffee filter
scissors
sand
jar
watch or clock with
  a second hand

Do the Numbers Add Up?

Here's an addition game you can play with your family on a long car trip.

1. Take turns picking a number from 1 to 20.
2. Look for a car's license plate whose numbers add up to the number that was picked.
3. Have a contest to see who can spot the license plate first.

Which numbers are the easiest to find?
Which numbers are the hardest?
Make up new rules for this game.
Teach your new game to a friend.